SEEING
THE
INVISIBLE
MATH

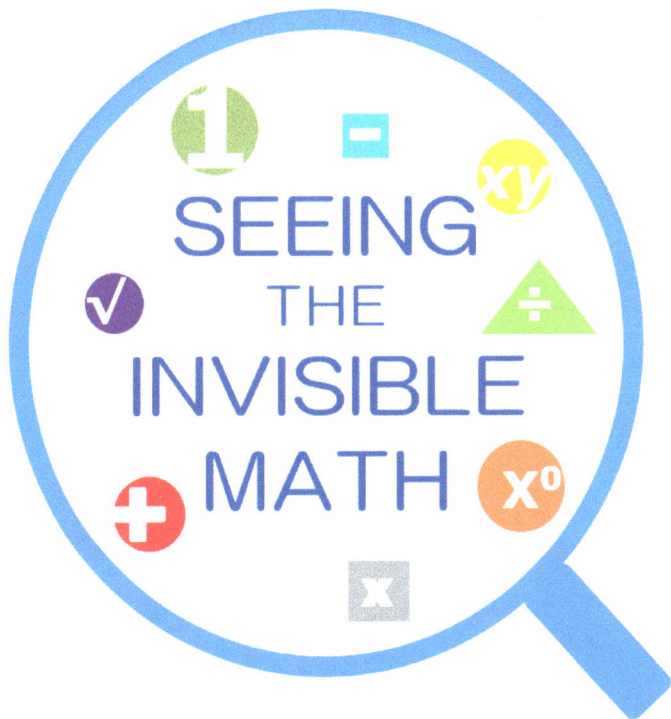

ARMIN BONIFACIO

First publication: 2021
Second publication: 2025

ISBN: 979-8-89860-559-9

Printed in the U.S.A.

www.toki.click

To those who get math, and those who don't (yet).

◻ Let's Get Started

Math can feel confusing when it looks like it's full of rules that just appear out of nowhere. But what if the real problem isn't you… it's that the clues were always there—just easy to miss without a guide.

This book is here to change that.

We'll uncover the invisible parts of math—the signs, numbers, and patterns that are always there, even when they're not written down. Once you see them, math won't feel like guessing anymore. It'll start to make sense.

⧉ Why This Matters

These hidden concepts show up everywhere: in solving equations, simplifying expressions, working with exponents, and more. They're not just random tips. They're *foundational skills* for advanced math.

When you know what's really going on behind the numbers, you'll make fewer mistakes and feel more confident in class, on tests, and in real life too.

📌 What to Expect

This book is organized into short, focused lessons.
Each concept is presented over three pages:

👀 A title and short idea that introduces what's
hiding

🔍 A clear explanation of the hidden truth and
a key point to keep in mind

🖊 A few quick problems to try on your own

✅ Answers and explanations to check your
thinking

Take your time. Look closely. Ask, *"What's really
going on here?"*

You're about to see math in a whole new way.

Let's begin.

TABLE OF CONTENTS

PART 1

The Power of 1

The Most Powerful Number

You Never See

INVISIBLE COEFFICIENT OF 1

 The Disappearing 1: Why x = 1 • x

🔍 The Hidden Truth:

Every variable has a number in front of it even if you don't see one.

When you write **x**, you're really writing **1 • x**. That invisible **1** is called the **coefficient**, and it tells you how many times **x** is being counted. We usually leave it out because it's so common, but it's always there.

This is important when combining like terms, solving equations, and factoring expressions.

✏️ Try It Yourself. What's Hiding?

Fill in the blanks to reveal what's really there:

1. **y**
   Rewrite with coefficient: _____ • y

2. **4y + z**
   Rewrite with all coefficients: 4 • y + _____ • z

3. **a + b + 5c**
   Rewrite with all coefficients:
   _____ • a + _____ • b + 5 • c

🍀 **Hint:** *If there's no number in front, it's not zero. It's just hiding.*

✅ Answers + Quick Explanation

Let's uncover the invisible numbers:

1. **1 • y**
→ There's an invisible 1 in front of any variable with no number.

2. **4 • y + 1 • z**
→ The 4 is written, but the z has a hidden +1.

3. **1 • a + 1 • b + 5 • c**
→ a and b each have an invisible +1 in front.

💡 Keep In Mind:

Recognizing the invisible **1** keeps you from making mistakes when combining like terms or solving equations. It's like the quiet teammate that holds the problem together.

INVISIBLE COEFFICIENT OF –1

👀 **The Minus with a Secret: Why –x = –1x**

🔍 The Hidden Truth:

Every negative variable has a hidden number too.

When you write **−x**, you're really writing **−1 • x**.
That invisible **−1** is still a **coefficient**. It tells you that you're working with the opposite of **x**. We often skip writing the **−1**, but it's always part of the expression.

This matters when distributing, factoring, or solving equations that involve negatives.

✏️ Try It Yourself. What's Hiding?

Fill in the blanks to reveal what's really there:

1. **−x + 3x**
   Rewrite with all coefficients shown:
   ___x + ___x

2. **−m − m**
   Rewrite with all coefficients shown:
   ___m + ___m

3. **−(a + b)**
   Rewrite with all coefficients shown:
   ___ • (a + b)

🍀 **Hint:** *That minus sign isn't floating alone. It's secretly −1.*

✅ Answers + Quick Explanation

Let's uncover the invisible numbers:

1. **−x + 3x → −1x + 3x**
→ The minus before x means −1 • x.

2. **−m − m → −1m + (−1m)**
→ You're adding two terms that each have an invisible −1 as their coefficient.

3. **−(a + b) → −1 • (a + b)**
→ That minus sign in front of parentheses is really a −1 being multiplied across.

💡 Keep In Mind:

Knowing there's a **−1** helps you distribute and simplify correctly. It's like spotting the hidden player changing the direction of the whole game.

INVISIBLE EXPONENT OF 1

 The Power You Don't See: Why $x = x^1$

🔎 The Hidden Truth:

Every variable you see, like **x** or **y**, is quietly carrying an exponent. You just don't write it because it's **1**.

In algebra, when no exponent is shown, it means the variable is being raised to the **first power**. That might not seem like much, but this rule becomes important when you're comparing exponents, simplifying expressions, or following exponent rules.

So when you see:
x

...it really means:
x^1

✏️ Try It Yourself. Reveal the Hidden Power.

Show the invisible exponent hiding in each of these:

1. $m + 4m^2$

   Rewrite with all exponents shown:

   ___ $+ 4m^2$

2. $x - y^3$

   Rewrite with all exponents shown:

   ___ $- y^3$

3. $a^2 \cdot a$

   Rewrite with all exponents shown and simplify using exponent rules. _____

   (Remember: when you multiply like bases, you add the exponents.)

✅ Answers + Quick Explanation

Let's reveal the exponents you didn't see:

1. $m + 4m^2 \rightarrow m^1 + 4m^2$
→ The m has an invisible exponent of 1. We usually don't write it, but it's there to show its power level.

2. $x - y^3 \rightarrow x^1 - y^3$
→ Same here, no written exponent means the exponent is 1.

3. $a^2 \cdot a \rightarrow a^2 \cdot a^1 = a^3$
→ When multiplying powers with the same base, add the exponents. So $2 + 1 = 3$.

💡 Keep In Mind:

Recognizing the invisible **1** exponent helps you correctly apply power rules in algebra and prevents errors when simplifying expressions or comparing terms.

4

ZERO EXPONENT RULE

👀 **The Power of Nothing: Why $x^0 = 1$**

🔎 The Hidden Truth:

This one surprises a lot of students—how can something raised to the **0** power equal **1**?

Any non-zero number or variable raised to the power of **0** is equal to **1**.

So:
$x^0 = 1$, $5^0 = 1$, and $(-3)^0 = 1$

Why? Because of how exponents work. Every time you divide matching powers, you subtract the exponents. And when you subtract a number from itself, you get zero.

For example:
$x^3 \div x^3 = x^{3-3} = x^0 = 1$

It's not magic. It's math.

✏️ Try It Yourself. Does It Equal 1?

Decide if each of these equals 1. Circle the ones that do, and rewrite them with explanations.

1. a^0
   Does it equal 1? _____
   Rewrite: _____

2. $x \cdot 1000^0$
   Does it equal 1? _____
   Rewrite: _____

3. 0^0
   Does it equal 1? _____
   Rewrite: _____

🍀 **Hint:** *Only non-zero numbers or variables raised to 0 equal 1.*

✅ Answers + Quick Explanation

Let's go one by one:

1. $a^0 = 1$
→ Any variable (as long as it's not 0) raised to the 0 power is 1.

2. $x \cdot 1000^0 = x \cdot 1 = x$
→ Just like above—any number raised to 0 equals 1. So when you multiply x by 1, the result is just x.

3. 0^0 = **undefined**
→ This is the tricky one. Zero raised to the 0 is **not** defined in math. It's a special case with no agreed answer, so you should not treat it as 1.

💡 Keep In Mind:

This rule helps simplify expressions and evaluate powers quickly. It also shows how exponents behave consistently even when the power seems like *"nothing."*

PART 2

Signs and Symbols
You Don't See

Why Symbols Matter Even the
Silent Ones

5

INVISIBLE POSITIVE SIGN

👀 **The Quiet Plus: Why 7 = +7**

🔍 **The Hidden Truth:**

In math, we only write the **plus sign (+)** when we really need to, but it's still there, silently doing its job.

When you see a number like **7**, it's actually **positive**. The **+** is just invisible.

So:
7 = +7
x = +x

This hidden sign helps when you're working with integers, solving equations, or comparing values. It also matters when you're lining up positive and negative numbers correctly.

🖊 Try It Yourself. Spot the Positive.

In each case below, write the number or variable with its invisible **+** sign included.

1. **4 • 7**
   Rewrite as: _____

2. **b**
   Rewrite as: _____

3. **2x + 3x + 5**
   Rewrite the expression to show all signs:

   _____

🍀 **Hint:** *If a number or variable has no sign in front, assume it's positive—even when combining like terms or constants.*

☑️ **Answers + Quick Explanation**

Let's highlight the invisible + signs:

1. **4 → (+4) • (+7)**
→ A plain number without a sign is actually positive.

2. **b → +b**
→ Just like with numbers, a lone variable is really positive unless marked otherwise.

3. **2x + 3x + 5 → (+2x) + (+3x) + (+5)**
→ All terms are positive. Writing the + signs shows they aren't neutral—they're positive.
When combined: (+2x) + (+3x) = 5x, so the final result is 5x + 5.

💡 Keep In Mind:

When working with integers and combining terms, knowing what's **positive** helps prevent sign mistakes. It also sets you up for success with absolute values, inequalities, and graphing.

IMPLIED MULTIPLICATION

 The Silent Operation:

Why 2x Means 2 • x

🔍 The Hidden Truth:

In algebra, multiplication happens more than you realize. It just hides in plain sight.

When a number and variable are side by side, like **2x**, there's an invisible • between them:

2x = 2 • x
3(x + 1) = 3 • (x + 1)
(x + 2)(x − 3) = (x + 2) • (x − 3)

This *"implied multiplication"* shows up everywhere in algebra. If you don't notice it, you might forget to apply the distributive property, make sign errors, or misread the expression.

✏️ Try It Yourself. Find the Hidden "•."

In each expression below, rewrite it by adding the invisible multiplication symbol (•).

1. **5x**
   Rewrite as: _____

2. **4(x + 2)**
   Rewrite as: _____

3. **(x + 3)(x – 1)**
   Rewrite as: _____

🍀 **Hint:** *When two expressions are next to each other, multiplication is happening even if you don't see the sign.*

✅ Answers + Quick Explanation

Let's insert the missing multiplication symbols:

1. **5x → 5 • x**
→ A number and a variable placed side by side mean multiplication.

2. **4(x + 2) → 4 • (x + 2)**
→ A number in front of parentheses means you multiply it by each term inside.

3. **(x + 3)(x − 1) → (x + 3) • (x − 1)**
→ Two grouped expressions placed next to each other indicate multiplication.

💡 Keep In Mind:

If you don't recognize where multiplication is hiding, you might skip the distributive property or try to combine things incorrectly. Spotting implied multiplication keeps your expressions accurate and clean.

NEGATIVE SIGN BEFORE PARENTHESES

👀 **The Hidden −1:**

Why −(x + 3) = −1 • (x + 3)

🔍 The Hidden Truth:

A minus sign in front of parentheses doesn't just sit there. It's multiplying the entire group by **–1**.

For example:
$-(x + 3) = -1 \cdot (x + 3)$

That means you need to multiply the **–1** to each term inside the parentheses.

This comes up all the time when solving equations, simplifying expressions, and factoring. If you forget the **–1**, you'll end up with sign errors.

✎ Try It Yourself. Don't Forget the –1.

Rewrite each expression by showing the invisible **–1**.

1. **–(a + b)**
   Rewrite as: _____

2. **–(3x – 5)**
   Rewrite as: _____

3. **Simplify –(x + 2) + 5**
   Step 1: Show the invisible –1
   Step 2: Distribute the –1
   Step 3: Write the simplified expression

☑️ **Answers + Quick Explanation**

Let's pull back the curtain:

1. **–(a + b) → –1 • (a + b)**
→ A minus in front = multiplying the group by –1.

2. **–(3x – 5) → –1 • (3x – 5)**
→ Again, we're applying the negative to everything inside.

3. **–(x + 2) + 5**
→ Step 1: **–1 • (x + 2) + 5**
→ Step 2: **–x – 2 + 5**
→ Step 3: **–x + 3**

💡 Keep In Mind:

A minus in front of parentheses isn't just decoration. It means everything inside is flipped. Forget to distribute the **−1**, and the whole solution can fall apart.

NEGATIVE SIGN BEFORE ABSOLUTE VALUE

The Outside Minus: Why −|x| = −1 • |x|

🔍 The Hidden Truth:

When a minus sign is placed **outside** an absolute value symbol, it means you're multiplying the result of the absolute value by **–1**.

Example:
–|4| = –1 • 4 = –4
–|–7| = –1 • 7 = –7

The absolute value always turns what's **inside** positive, but the minus sign **outside** will flip the result afterward. That negative is not part of the absolute value. It's waiting outside, ready to act.

✏️ Try It Yourself. Flip It After Absolute Value.

Evaluate or rewrite each expression. Pay attention to the minus sign's position.

1. $-|9|$
   Answer: _____

2. $-|-5|$
   Answer: _____

3. $-|a|$
   Answer: _____

🍀 **Hint:** *The minus is not inside the bars. It comes after absolute value has done its job.*

✅ **Answers + Quick Explanation**

Let's see what that outer minus was really doing:

1. **$-|9| = -1 \cdot 9 = -9$**
→ 9 is already positive, so the negative flips the result.

2. **$-|{-5}| = -1 \cdot 5 = -5$**
→ The absolute value made –5 into 5. Then the minus sign turned it negative again.

3. **$-|a| = -1 \cdot |a| = -1a = -a$**
→ That minus stays outside. It multiplies whatever comes out of the absolute value bars.

💡 Keep In Mind:

If you think the negative is part of the absolute value, you'll get the wrong sign. This rule clears up one of the most common algebra mistakes.

9

NEGATIVE EXPONENT
EQUALS RECIPROCAL

Flip the Power: Why $x^{-1} = 1 \cdot \dfrac{1}{x}$

🔍 The Hidden Truth:

When an **exponent** is **negative**, it doesn't mean the number itself is negative. It means you **flip** it.

So:

$$x^{-1} = \frac{1}{x}$$

$$5^{-2} = \frac{1}{5^2} = \frac{1}{25}$$

$$(ab)^{-3} = \frac{1}{(ab)^3} = \frac{1}{a^3b^3}$$

Negative exponents mean *"put this on the other side of the fraction bar."* That's why x^{-1} ends up in the denominator. This shows up a lot in simplifying expressions, scientific notation, and rational exponents.

✏️ Try It Yourself. Flip the Base.

Rewrite each expression using the rule for negative exponents.

1. x^{-2}

   Rewrite as: _____

2. 3^{-1}

   Rewrite as: _____

3. $\dfrac{1}{(xy)^{-10}}$

   Rewrite as: _____

✳️ **Hint:** *Negative exponent = reciprocal. Flip the base and make the exponent positive.*

✅ Answers + Quick Explanation

Let's flip these expressions the right way:

1. $x^{-2} = \dfrac{1}{x^2}$

→ The negative tells you to move the base to the denominator.

2. $3^{-1} = \dfrac{1}{3}$

→ Simple reciprocal.

3. $\dfrac{1}{(xy)^{-10}} = \dfrac{(xy)^{10}}{1}$

→ Flip the base to the top and make the exponent positive. Then apply the exponent to both variables (see *Concept 15*, page 79). The final answer is $x^{10}y^{10}$.

💡 Keep In Mind:

Many students confuse a **negative exponent** with a negative value of the expression, but they're totally different. A negative exponent tells you to **flip** the base—not that the answer will be negative. Understanding this invisible flip helps simplify and solve exponential expressions correctly.

PART 3

Numbers in Disguise

How Numbers Hide

in Plain Sight

THAT DOT NEVER DISAPPEARS. IT'S ONLY HIDING!

5=5.0

10

INVISIBLE DECIMAL POINT

AND ZEROS

👀 **The Hidden Dot: Why 5 = 5.0 = 5.00...**

🔎 The Hidden Truth:

Every whole number is secretly a **decimal**. It just doesn't show it.

5 = 5.0 = 5.00 = 5.00000...

This invisible decimal lets us switch smoothly between whole numbers and decimal numbers when comparing, adding, or subtracting. It's also what makes aligning digits in place value or operations like long division possible.

✏️ Try It Yourself. Find the Dot.

Rewrite each whole number as a decimal, then add more zeros if needed.

1. **7**

   Write as a decimal with one zero: _____

2. **12**

   Write as a decimal with two zeros: _____

3. **3 + 0.4**

   Rewrite 3 as a decimal to help line it up:

   _____

🍀 **Hint:** *Zeros after the decimal don't change the value. They just show precision.*

✅ Answers + Quick Explanation

Here's what the invisible dots reveal:

1. **7 → 7.0**
   ➡ Same value, now in decimal form.

2. **12 → 12.00**
   ➡ Still 12, but now with two decimal places.

3. **3 + 0.4 → 3.0 + 0.4**
   ➡ Writing 3 as 3.0 helps line up the decimals for addition.

💡 Keep In Mind:

The hidden decimal lets us write whole numbers in decimal form—like **3** as **3.0**—so we can line up digits correctly when adding, subtracting, or comparing values. It helps avoid place value mistakes and keeps decimal operations clean and accurate.

11

INVISIBLE FRACTION FORM

The Fraction You Never See:

$$\text{Why } 5 = \frac{5}{1}$$

🔍 **The Hidden Truth:**

Every whole number can be written as a **fraction**. It just has an invisible **1** on the bottom.

So:

$9 = 9/1$ or $\dfrac{9}{1}$

$x = x/1$ or $\dfrac{x}{1}$

This is especially useful when you're multiplying or dividing with other fractions, because it helps you treat whole numbers and fractions the same way.

✏️ Try It Yourself. Make It a Fraction.

Rewrite each whole number or variable as a fraction over 1.

1. **8**
   Rewrite as: _____

2. **a**
   Rewrite as: _____

3. **4 ÷ 2/3**
   Rewrite 4 as a fraction and solve:

   _____

🍀 **Hint:** *Any number = itself over 1.*

✅ Answers + Quick Explanation

Let's turn those into fractions:

1. $8 \rightarrow \dfrac{8}{1}$

   → Now it behaves like a fraction in operations.

2. $a \rightarrow \dfrac{a}{1}$

   → Variables can be written over 1 too!

3. $\dfrac{4}{1} \div \dfrac{2}{3} = \dfrac{4}{1} \cdot \dfrac{3}{2} = \dfrac{12}{2} = 6$

   → Writing 4 as 4/1 lets you multiply by the reciprocal of the second fraction.

💡 Keep In Mind:

Turning whole numbers into fractions keeps your operations consistent. It's a simple move that can prevent mistakes when multiplying or dividing by other fractions.

12

INVISIBLE RADICAL INDEX OF 2

👀 **The Unspoken Root: Why $\sqrt{x} = \sqrt[2]{x}$**

🔍 The Hidden Truth:

When you see the radical symbol $\sqrt{x}$, it's actually a **square root**, even though there's no number written.

That's because the index (the little number above the $\sqrt{}$) is **2** by default. We don't write it, but it's always there.

So:
$$\sqrt{x} = \sqrt[2]{x}$$

If the root is **not** square (like cube root or fourth root), we must write the index. But for square roots, the 2 is just... invisible.

✏️ Try It Yourself. Name the Index.

Write the full root expression by filling in the missing index.

1. $\sqrt{25}$
   Rewrite with index: _____

2. $\sqrt{x}$
   Rewrite with index: _____

3. $\sqrt[3]{8}$
   Which root is this? Is the index invisible?

   _____

🍀 **Hint:** *Only square roots get the invisible 2. Every other root must show its index.*

✅ Answers + Quick Explanation

Let's make those hidden 2s show up:

1. $\sqrt{25} \rightarrow \sqrt[2]{25}$
→ This means *"What number times itself equals 25?"* Answer: 5.

2. $\sqrt{x} \rightarrow \sqrt[2]{x}$
→ Even variables follow this rule.

3. $\sqrt[3]{8} \rightarrow$ cube root of 8; index is not invisible.
→ You must write the 3 for cube root. Only square roots skip it.

💡 Keep In Mind:

If you don't realize square roots have an invisible **2**, you might apply the wrong exponent rule or confuse root types. This clears it up.

13

INVISIBLE DENOMINATOR

IN ROOTS

👀 **From Root to Power: Why $\sqrt{x} = x^{1/2}$**

🔍 The Hidden Truth:

Every root can also be written as an exponent, and that exponent is a **fraction**.

Let's look at $\sqrt{x^3}$. It means "the square root of x^3."

Even though the index **2** isn't written, it's still there.

So:

$\sqrt{x^3} = x^{3/2}$

The **2** is the invisible denominator, and the **3**, the exponent on **x**, is the numerator of the fraction.

✏️ Try It Yourself. Rewrite the Roots.

Change each root expression into a power with a fractional exponent.

1. $\sqrt{x}$
   Rewrite as: _____

2. $\sqrt[3]{y}$
   Rewrite as: _____

3. $\sqrt{z^2}$
   Rewrite as: _____

🍀 **Hint:** *The index of the root becomes the denominator of the exponent. The power inside the root becomes the numerator.*

✅ Answers + Quick Explanation

Let's turn roots into powers:

1. $\sqrt{x} \to x^{1/2}$
→ Square root = ½ exponent. Remember, x has an invisible exponent of 1, which becomes the numerator.

2. $\sqrt[3]{y} \to y^{1/3}$
→ Cube root = ⅓ exponent.

3. $\sqrt{z^2} \to z^{2/2} = z^1 = z$
→ Power inside becomes the numerator; root index is 2 (invisible). The exponent $^2/_2$ simplifies to 1.

💡 Keep In Mind:

Fractional exponents and radicals are two sides of the same coin. Seeing the connection helps you simplify or solve complex expressions with more flexibility.

14.

INVISIBLE POWER RULES IN SCIENTIFIC NOTATION

Power Shift: Why 5 × 10³ = 5000

🔍 The Hidden Truth:

Scientific notation uses powers of **10** to shift place value, but that shift is often left unexplained.

So when you see:
5 × 10³,

what it really means is:
Move the decimal 3 places to the right → 5000

The power tells you how many zeros (or spaces) to move the decimal—so **10³** means three places to the right.

✏️ Try It Yourself. Show the Shift.

Convert each scientific notation to standard form.

1. **4×10^2**
   Answer: _____

2. **6.7×10^1**
   Answer: _____

3. **3×10^0**
   Answer: _____

🍀 **Hint:** *$10^0 = 1$. The exponent tells you how many places to move the decimal.*

✅ Answers + Quick Explanation

Here's where those powers took us:

1. **$4 \times 10^2 = 400$**
➡ Decimal moves 2 places right.

2. **$6.7 \times 10^1 = 67$**
➡ One place right.

3. **$3 \times 10^0 = 3 \times 1 = 3$**
➡ Any number to the 0 power = 1, so no shift needed.

💡 Keep In Mind:

Understanding how powers of **10** change place value is the foundation for scientific notation, metric conversions, and mental math. It's simple, but powerful.

PART 4

Grouping and Order

The Math Behind the Math

15

IMPLIED PARENTHESES IN ORDER OF OPERATIONS

👀 **Where the Power Goes:**

Why $3x^2 \neq (3x)^2$

🔍 The Hidden Truth:

When you see something like **$3x^2$**, the exponent only applies to the **x**, not the **3**.

So:
$3x^2 = 3 \cdot (x^2)$

But:
$(3x)^2 = (3^2) \cdot (x^2) = 9x^2$

Because **$(3x)^2$** has parentheses, the exponent **2** applies to **both** the 3 and the variable **x**.

The difference is whether the whole thing is squared, or just the variable. If you don't see parentheses, don't assume they're there.

✏️ Try It Yourself. Which One's Squared?

Decide which part the exponent is affecting. Rewrite with parentheses to show what's actually happening.

1. **$5x^2$**

   Rewrite as: _____

2. **$(2x)^2$**

   Simplify it: _____

3. **$-x^2$**

   Is the negative sign included in the square?

   Yes / No

   Rewrite as: _____

🍀 **Hint:** *Only what's directly attached to the exponent is affected. If there are no parentheses, the exponent does not apply to everything.*

✅ Answers + Quick Explanation

Let's clarify what's squared and what's not:

1. $5x^2 \rightarrow (5) \cdot (x^2)$
→ The 5 is not being squared—only the x is.

2. $(2x)^2 = (2^2) \cdot (x^2) = 4x^2$
→ Both the 2 and the x are squared.

3. $-x^2 = -(x^2) = -x^2$
→ The exponent 2 applies only to x. The negative sign stays as it is.

💡 Keep In Mind:

Assuming the wrong grouping leads to totally different results. Being clear about what's squared protects you from major mistakes.

16

IMPLIED GROUPING BY
FRACTION BAR

👀 **The Great Divider:**

Why 1/(x + 2) ≠ 1/x + 2

🔎 The Hidden Truth:

When you write a fraction like $\dfrac{1}{x+2}$, the fraction bar **groups** the entire denominator even if parentheses aren't written.

So:

$\dfrac{1}{x+2}$ means the whole **x + 2** is the denominator.

But:

1/x + 2 actually means **(1 / x) + 2** or $\dfrac{1}{x} + 2$ which is totally different. Only **x** is in the denominator of **1**.

The fraction bar acts like invisible parentheses around the entire numerator and denominator.

✏️ Try It Yourself. What's Grouped?

Choose which part of each expression is in the denominator. Add parentheses to show grouping clearly.

1. **1 / x + 3**

   Rewrite as: _____

2. $\dfrac{1}{x+3}$

   Is the entire denominator grouped? Yes / No

3. $\dfrac{x+2}{y-1}$

   What's in the numerator? _____

   What's in the denominator? _____

🍀 **Hint:** *The bar separates top and bottom completely, even if you don't see parentheses.*

✅ Answers + Quick Explanation

Let's make the groupings visible:

1. **1 / x + 3 → (1 / x) + 3**
→ Only x is in the denominator of 1.

2. **1 / (x + 3) → Yes**
→ Entire x + 3 is grouped by the fraction bar.

3. **(x + 2) / (y – 1)**
→ Numerator: x + 2
→ Denominator: y – 1
→ Parentheses help remove confusion.

💡 Keep In Mind:

Misreading groupings in a fraction changes the entire meaning. Understanding this invisible grouping rule helps prevent the most common mistake students make in rational expressions.

INVISIBLE ZERO CONSTANT

👀 **The Silent Tagalong in Polynomials:**

Why $-3x^2 + 2x$ is really $-3x^2 + 2x + 0$

🔍 The Hidden Truth

When a constant term is missing in a polynomial, it doesn't mean it's gone—it means it's **zero**. That **0** might not be written, but it plays a role when you're solving, graphing, or analyzing expressions.

For example:
$-3x^2 + 2x$ is really $-3x^2 + 2x + 0$

✏️ Try It Yourself. Add the Missing Constants.

Write each expression again by including the invisible constant at the end.

1. $x^2 + x \rightarrow$ _____

2. $-4x^3 + 9x \rightarrow$ _____

3. $5x^2 - 2x^4 \rightarrow$ _____

🍀 **Hint**: If there's no constant term, just add +0 to show it's still there.

✅ **Answers + Quick Explanation**

Let's complete the expressions by showing the constant:

1. $x^2 + x \rightarrow x^2 + x + 0$

   → The constant isn't written, but it's really 0.

2. $-4x^3 + 9x \rightarrow -4x^3 + 9x + 0$

   → Including the 0 keeps the format complete and useful in many steps.

3. $5x^2 - 2x^4 \rightarrow -2x^4 + 5x^2 + 0$

   → The constant is still missing even when the terms are out of order. Rewriting in standard form and adding 0 makes it clearer.

💡 Keep In Mind

Knowing that the constant is **0** can help you:

• **Graph more accurately**. The y-intercept is (0, 0), so the graph passes through the origin.

• **Stay organized**. When factoring, dividing, or lining up like terms, a full set of terms (including the constant) keeps your math clean and error-free.

18

INVISIBLE ZERO COEFFICIENTS

👀 **The Missing Terms That Are Really 0s:**

Why $3x^2 + 5 = 3x^2 + 0x + 5$

🔍 The Hidden Truth

When a power of **x** is missing in a polynomial, it means its coefficient is **0**.

We usually skip writing the missing power, but putting it in helps when you rewrite terms from highest to lowest. It's especially helpful for steps like polynomial division or using formulas that need every power to show up.

For example:

$3x^2 + 5$ is really **$3x^2 + 0x + 5$**

$4x^3 - 7$ is really **$4x^3 + 0x^2 + 0x - 7$**

$2x^4 + 2x^2 + 1$ is really **$2x^4 + 0x^3 + 2x^2 + 0x + 1$**

✏️ Try It Yourself. Fill in the Zeros.

Rewrite each expression by inserting the invisible zero coefficients.

1. $4x^3 - 7$

   ➡️ _____

2. $7x^5 + x + 10$

   ➡️ _____

3. $2x^4 + x^2 - 3x$

   ➡️ _____

Hint: Add a zero term for each missing power of x, starting from the highest down to the constant.

✅ Answers + Quick Explanation

1. $4x^3 - 7 \rightarrow 4x^3 + \underline{0x^2 + 0x} - 7$

   ➡ We include x^2 and x terms with coefficients of 0 to show all degrees from 3 down to 0.

2. $7x^5 + x + 10 \rightarrow 7x^5 + \underline{0x^4 + 0x^3 + 0x^2} + x + 10$

   ➡ x^4, x^3, and x^2 terms were missing, so we add $0x^4$, $0x^3$, and $0x^2$ to show every degree from 5 down to 0.

3. $2x^4 + x^2 - 3x \rightarrow 2x^4 + \underline{0x^3} + x^2 - 3x + \underline{0}$

   ➡ x^3 and the constant term were missing, so we add $0x^3$ and 0.

💡 Keep In Mind:

Using invisible **zero** coefficients helps keep expressions in **standard form** and avoids mistakes when aligning terms or applying structured steps. It keeps the math clean and complete.

After *Seeing the Invisible Math...*

Visit **www.toki.click** to explore more titles in *The Life Skills Playbook Series for Teens and Young Adults*—and discover tools to help you grow, reflect, and move forward, one moment at a time.

About tokï

Tokï is a word from Japanese that means *"time"* or *"moment."*

And that's what the journey of life is made of—little moments that shape who we become.

From wonder to wisdom, **tokï** captures the spirit of growing up—embracing curiosity, facing challenges, and becoming stronger with every step.

tokï isn't just a name.

It's a vibe.

It's time well lived.